STATE MACHINES IN VHDL
Multipliers
Vol. 2.1

State Machine Design for Arithmetic Processes

Daryl Ray Hawkins

Copyright © 2013 Daryl Ray Hawkins
All rights reserved.
ISBN: 1493690531
ISBN-13: 978-1493690534

Revision Dates
December 2013 First Release

December 2014 Revised

- Textual and code changes regarding the Sticky-Bit, some code changes from unsigned signals to signed.

Table of Contents

Overview 4
1   Prerequisites 4
2   Fixed Versus Floating-Point 5
3   Limiting Factors 7
    3.1   Carry propagation 7
    3.2   Integer Size 7
4   Normalizing, Rounding, and Overflow 9
    4.1   Fixed-Point 10
    4.2   Floating-Point 11
    4.3   Rounding Modes 12
    4.4   The Sticky Bit 15
5   Simple Sequential Multiplier 16
6   Booth Sequential Multiplier 24
7   Combined Partial Product Multiplier 32
8   Combined Array Multiplier 43
9   Other information on Multipliers 67
10   Addendum 68

Overview

Multiplication is the next most common arithmetic operation after addition and subtraction. Many approaches are available to the designer, each with its own pros and cons. Speed versus resource use are generally the competing priorities, with power consumption in the mix.

Provided in this book are four implementation. Complexity and performance increase with each successive design. Each are done in such a way to balance resource use with the maximum obtainable clock speed.

Suggestions for further improving performance are provided at the end of each chapter. While simplicity is usually best, there are cases where raw performance is an imperative and should be employed when needed. Section *3.1 Carry Propagation* provides additional direction on increasing performance, namely through fast adders.

As a reminder to the reader: because of the gate resources available in FPGA and ASIC devices, clever arrangements within a design can yield respectable performance.

An example for each design, fully implemented as a state machine, is also provided at the end of each chapter.

Copies of all source code used in this book can be acquired through the web site listed below -- under *publications*, or *pubs*.

http://www.hawkinseng.com

1 Prerequisites

State Machine techniques used throughout this book are covered in the "STATE MACHINES IN VHDL *Composition* Vol. 1" book, which is a prerequisite, and should be reviewed by the reader.

Additionally, reviewing the IEEE Std 754-1985/2008 Floating-Point specifications is also recommended.

2 Fixed Versus Floating-Point

All operations are inherently fixed-point and are favored over floating-point, primarily because of the additional overhead burden associated with floating-point normalization (required implied 1 format). The down side to fixed-point is the reduced accuracy when approaching the lower boundaries, and the magnitude confinements of its explicit range.

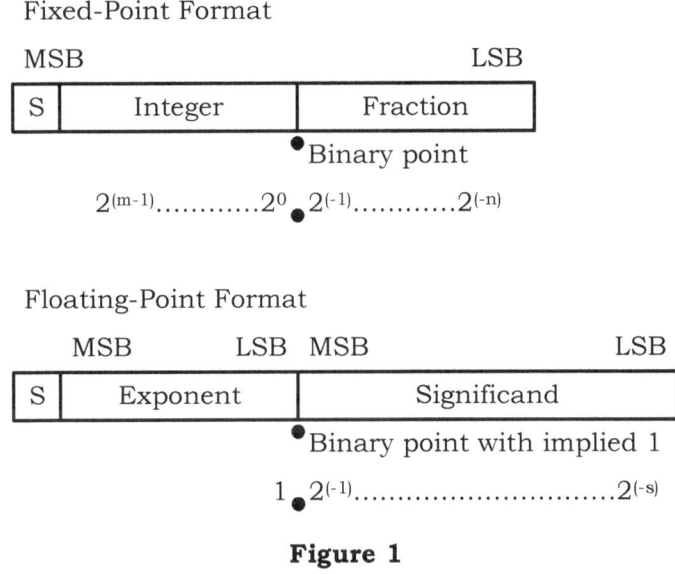

Figure 1

Fixed-point is a two's complement binary number with a fixed number of integer bits and a fixed number of fractional bits with the binary point being implicit between the two. The industry Q notation for fixed-point binary numbers is Qm.n, 'm' representing the number of integer bits excluding the sign bit; 'n' representing the number of fractional bits.

Note: *An example for Qm.n would read Q15.16 for a 32-bits.*

Floating-point is a signed magnitude number with an implied binary point. The significand (all bits to the right of the binary point) is left-shifted so that the most significant '1' bit is positioned to the left of the implied binary point and discarded (not saved), while offsetting the exponent (bias) accordingly.

Two of the most common implementations are single and double precision, 23-bit and 52-bit significand (mantissa) excluding the implied 1; 24-bit and 53-bit including it.

The number of bits allocated for the exponent is 8-bits for single precision and 11-bits for double. Their nominal value is an offset from 0, which is +127 and +1023, respectively. Because they are offset values instead of absolute values, they are referred to as a bias instead of an exponent; sometimes called an exponent bias.

Associated extended formats are not needed with multiplication. Extended means that more bits are used during the operational stage than are kept after normalization.

> Note: *All implementations in this book are done in fixed-point. Where warranted, notes are provided on floating-point for consideration by the designer.*

3 Limiting Factors

3.1 Carry propagation

Carry propagation in arithmetic design is one of the greatest limiters of a design's maximum clock speed. Fast adders are a fundamental requirement for performance.

High performance FPGA devices employ embedded carry-chains within the silicon that run through each vertical column of logic blocks. These are used by the programmable logic for creating fast adders. Synthesis tools take advantage of these chains when targeting such devices.

Other FPGA or ASIC devices employ tile or gate array resources and do not have embedded carry functions. In those cases, the designer must explicitly design their own fast adder. Listed below are some of the more commonly known techniques. Each require detailed design and will use more logic resources.

- Carry Save Adders
- Carry Look-ahead Adders
- Carry Select Adders
- Carry Skip Adders

Even more design effort is required when implementing ultra fast adders. Compressors, adder trees, and hybrids along with recoding schemes, are a few of the more elaborate techniques and can be used if necessary. Reference material on these and other similar schemes are available in textbooks and in published white papers.

If the latter is not practical, pipelining and breaking up adders into multiple stages provide the greatest returns.

3.2 Integer Size

There is no inherent limit to the size of an adder in VHDL. There are, however, limitations in representing and monitoring large fixed-point numbers in source code.

Generally, designers use conversion functions to convert between real data objects and binary numbers. Real signals or variables can be expressed textually in the floating-point style. This is useful when assigning constants in the source code and for monitoring objects during simulation.

The exponent (**) operator and CONV_INTEGER or to_integer functions are limited in range between plus and minus 2147483647 (2147483648), reducing the conversion range of the integer portion to 32-bits (sign included) and the fractional portion to 31-bits (no sign), each.

> Note: *Provided in the addendum are functions that will convert from fixed-point to real (simulation floating-point data objects) that extend past the integer limitation of VHDL. They may also be used to assign constants with floating-point text representation.*

4 Normalizing, Rounding, and Overflow

The product size in bits is twice that of the operand size, which are the multiplier and multiplicand, if they are of equal size. Whether fixed-point or floating-point, the initial product value must be re-normalized to fit the original operand size. The remaining bits are not simply discarded, but instead, are used to detect overflow and compute rounding.

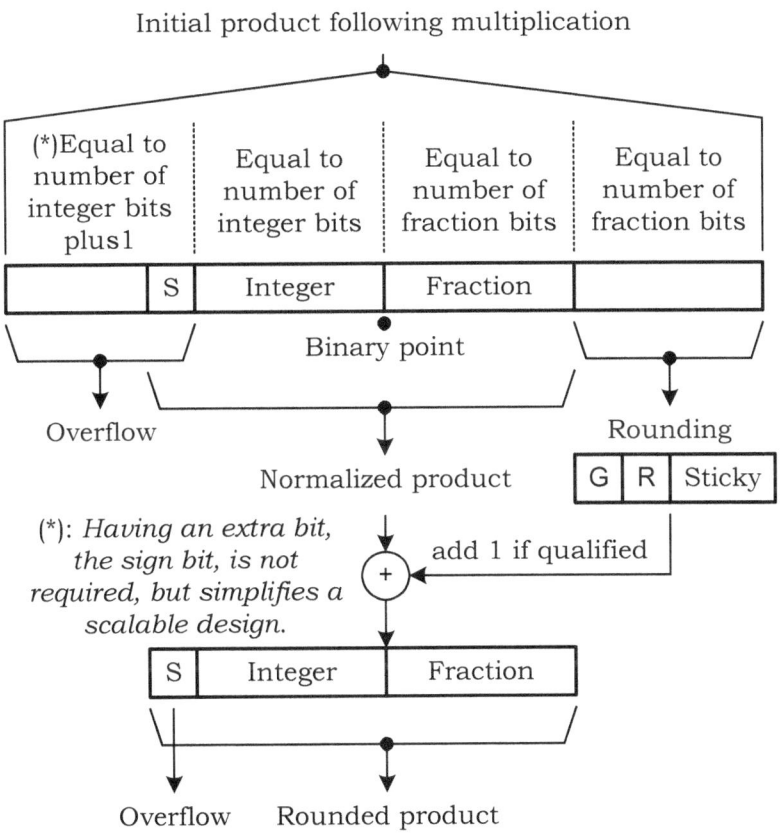

Figure 2 Fixed-Point Normalizing

Overflow with fixed-point is dependent on where the normalized binary-point is relative to the initial product (bits above the sign bit). In contrast, floating-point does not overflow unless the exponent reaches its upper bounds.

Rounding is more than the 0.5 principle taught in grammar school. In fact, there is a dedicated science to rounding numbers. Accumulative errors due to recursive multiplies can significantly effect the outcome of computational intensive processes.

One of the most common rounding modes is referred to as *Round-To-Nearest-Even*. This implementation uses three extra bits: guard, round, and sticky, or *GRS*. This paradigm is used in examples provided in this book, except where otherwise noted. However, the designer should review the implications of all standard rounding modes for their particular application. A list of rounding modes is provided at the end of this chapter.

4.1 Fixed-Point

Figure 2 illustrates the steps for finalizing the product of a fixed-point multiplication. First is the format of the initial product, in its entirety, following the multiplication. The binary-point position of the normalized product is offset from the right by two-times the number of fractional bits. Those bits below or to the right of the fraction bit field are used for rounding (along with the LSB of the normalized product), which are also equal to the number of fractional bits. Above the integer bit field, including the sign, are bits used for detecting overflow.

> Note: *the operand format used must have at least three fractional bits in order to implement the GRS rounding scheme. A small number of fractional bits can cause large round-up-in-magnitude errors. Disabling rounding, which simply truncates, may be better when implementing operands with a small number of fraction bits.*

1). Initial product size = 1+ (2 x Integer bits) + (2 x Fraction bits)
2). Normalized binary-point position = (2 x Fraction bits)

> Note: *the sign bit is not required in computing the initial product size unless signed multiplication is use -- how the polarity of the overflow bits are interpreted is based on the sign of the product.*

4.2 Floating-Point

Figure 3, in contrast, shows the steps for finalizing the product of a floating-point multiplication.

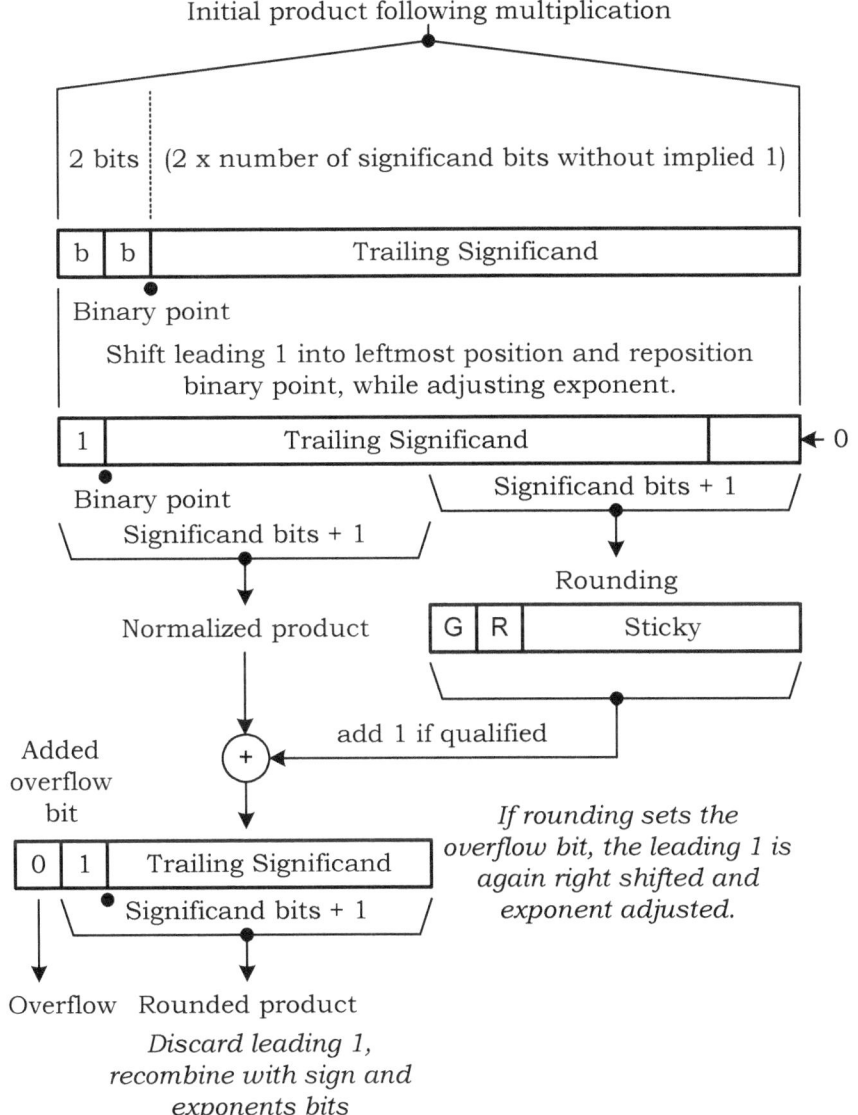

Figure 3 Floating-Point Normalizing

At the top is the format of the initial floating-point product following the multiplication. Unlike fixed-point, all floating-point operands are signed-magnitude, meaning the significand portion of the operand is always a positive binary number, thus only unsigned multiplication is required.

The first step is to position the leading 1 of the significand to the left and the virtual binary-point to its right, shifting in zeros, then adjusting the exponent value accordingly.

> Note: *when multiplying two floating-point numbers, their corresponding exponents are added. Further adjustments to the exponent occur during the normalization and rounding steps.*

Next, the bits in the lower half of the product field, along with the LSB of the normalized product, are used to qualify rounding. If qualified, 1 is added to the normalized product, and afterwards the leading 1 is again repositioned to the left of the virtual binary-point. In the diagram an extra bit is added to the left, in the event the rounding step causes an overflow. This only qualifies as an error if the exponent is at its upper bound.

Afterwards, the rounded product is stripped of its leading '1', making it implied. Then the exponent and sign are recombined with the significand into the floating-point format.

Round To Nearest Even

Guard – the MSB, most left, of the rounding bits.
Round – one bit to the right of the Guard bit.
Sticky – An OR of all the bits, see section *4.4 The Sticky Bit*.

GRS > "100" or (GRS = "100" and LSB of normalized product = '1')

4.3 Rounding Modes

Some fixed-point and all floating-point implementations represent the significand as a positive binary number, denoting the polarity with a sign bit -- sometimes referred to as sign-magnitude. As a result, rounding has the same effect on positive and negative numbers, because all products are rounded as positive numbers. In contrast, when using two's complement numbering, positive and negative numbers may require different treatment, depending on the rounding mode used. Rules are as follows:

Positive numbers – Truncation rounds the number down, round-down, making it less positive or towards zero; whereas adding 1 rounds the number up, round-up, more positive or towards + infinity or away from zero. If there are additional ones below the normalized result, the number is larger.

Negative numbers – Truncating rounds the number down, round-down, making it more negative or towards (-) infinity or away from zero; whereas adding 1 rounds the number up, round-up, less negative or towards zero. If there are additional ones below the normalized result, the number is smaller.

Below is a list rounding modes:

- ***Round toward zero (IEEE)***
[+ infinity] (positives) -> 0 <- (negatives) [-Infinity]

Truncate positive numbers, round-down. If any lower fractional bits of a negative numbers are '1', round-up.

- ***Round away from zero***
- ***Round toward infinity*** or ***Up-magnitude***
[+ infinity] <- (positives) -- 0 -- (negatives) -> [-Infinity]

For positive numbers, round-up if any lower fractional bits are '1'. Truncate negative numbers, round-down.

- ***Round toward positive (+) infinity (IEEE)***
[+ infinity] <- (positives) -- 0 <- (negatives) [-Infinity]

If any lower fractional bits are '1' for either positive or negative numbers, round-up.

- ***Round toward negative (–) infinity (IEEE)*** or ***Truncation***
[+ infinity] (positives) -> 0 -- (negatives) -> [-Infinity]

Truncate both positive and negative numbers, round-down

- ***Round toward nearest even (IEEE default)***
- ***Convergent***

STATE MACHINES IN VHDL *Multipliers* Vol. 2.1

Rounding to nearest has several sub-categories. The method referred to as Bankers is chosen here.

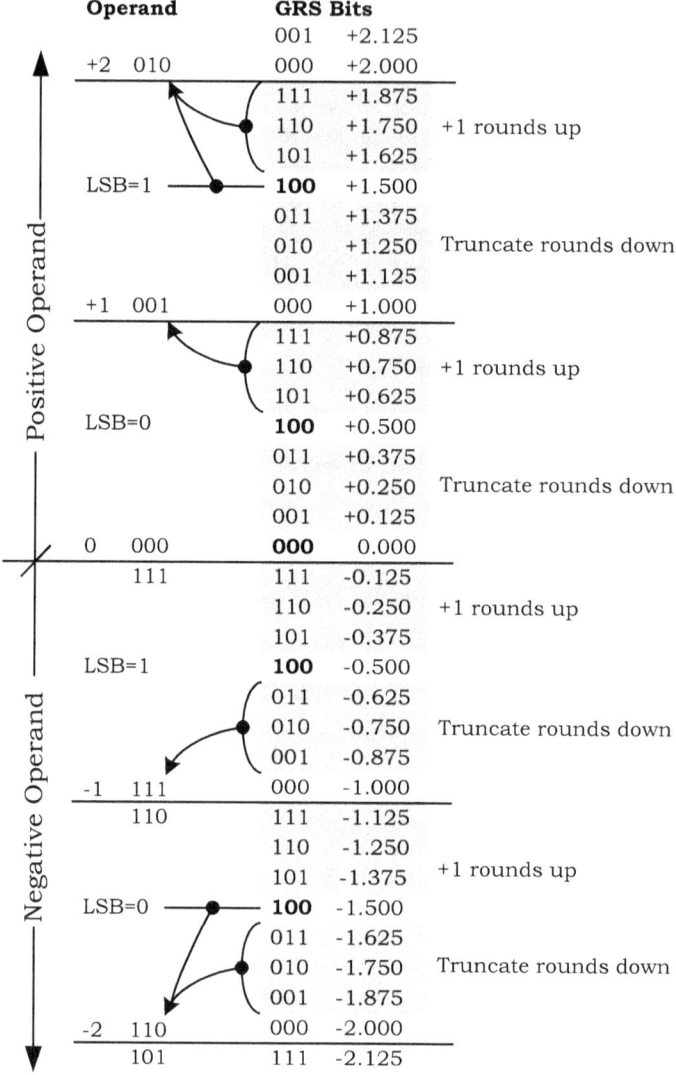

Figure 4

When using this mode on floating-point numbers, which operate on positive operands only, the result will effectively round away from zero or toward infinity. Since positive and negative operands are only different because of the sign bit, the rounding bias is the same for both positive and negative operands. This is what most designer's want.

Likewise, when using this mode on two's complement fixed-point numbers, the rounding bias appears to be symmetrical, as illustrated in Figure 4 and the qualifiers below.

   Positive Operands

      GRS > "100" or (GRS = "100" and LSB = '1')   add 1

      Otherwise truncate

   Negative Operands

      GRS < "100" or (GRS = "100" and LSB = '0')   truncate

      Otherwise add 1

   Combining the above qualifiers surprisingly reduces to:

      GRS > "100" or (GRS = "100" and LSB = '1')   add 1

      Otherwise truncate

While some experts may challenge this last point on directional rounding bias and symmetry, it is the implementation of choice throughout this book.

4.4 The Sticky Bit

The Sticky-Bit represents an OR of any bits to the right of, or less significant to, the round bit. Its value depends on the numbering representation used.

If the product is either a sign-magnitude or a positive two's complement number, sticky is set to '1' if any of the lower bits are '1', else it is set to '0'. If the product is a negative two's complement number, sticky is set to '0' if any of the lower bits are '0', otherwise it is set to '1'.

5 Simple Sequential Multiplier

The *Simple Sequential Multiplier* is simple and very efficient. While easy to implement, it has intrinsic limitations. It supports only positive operands and has a lengthy execution time.

Sign conversion of negative operands prior to multiplication and possibly the resulting product are required. Execution time, or the number of clock cycles to produce the result, is equal to number of bits in the operand plus additional overhead for sign conversions and rounding.

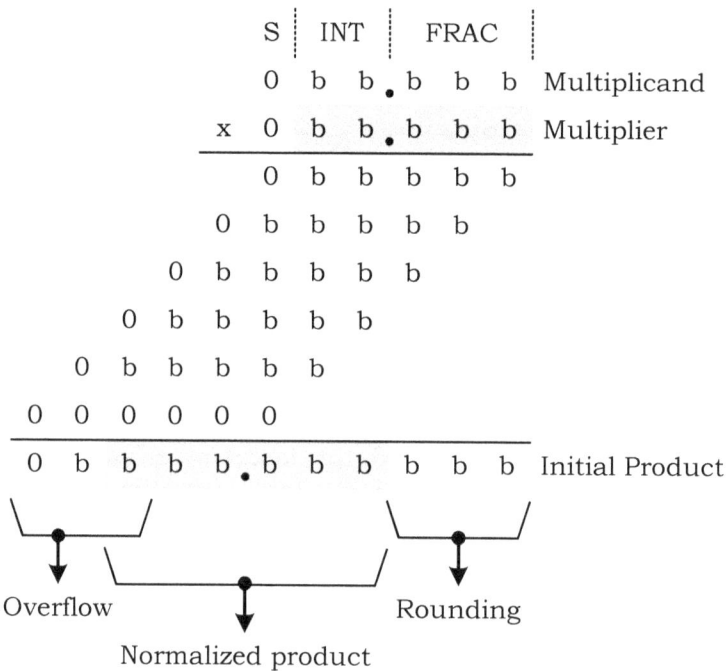

Figure 5

Note: *since this example only multiplies positive numbers, the sign bit need not be included in the sequence. However, leaving it simplifies the state machine design.*

Figure 5 depicts multiplying two positive fixed-point numbers using the pencil-and-paper approach. If the multiplier bit is a '1' then the multiplicand value is repeated, shifted to the left by one bit. If a '0', then a zero value is inserted instead. The initial product is the collective sum of each successive entry.

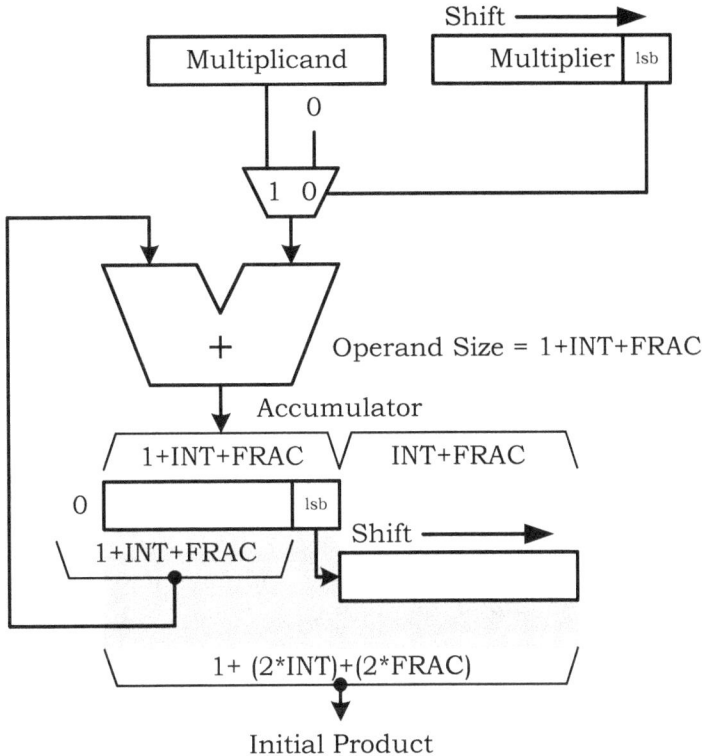

Figure 6

Figure 6 shows the structural design required to implement the *Simple Sequential Multiplier*. On every clock cycle, the LSB of the multiplier operand is evaluated then right-shifted, bringing the next bit into position for the next clock cycle. If the LSB bit is '1', the multiplicand is added to the upper half of the accumulator (offset by 1, thereby truncating the LSB), otherwise, zero is added.

As the upper half of the accumulator is updated with the sum from the adder, the LSB from the previous sum is right-shifted into the lower half of the accumulator as the entire lower portion is right shifted.

The number of clock cycles required to complete the operation is equal to the number of bits in the multiplier, LSB through MSB, resulting in the initial product being present in the combined upper and lower accumulator.

Note: *see chapter 4 for Normalizing, Rounding, and Overflow.*

Optional Improvements

Since the operands are positive, the sign bit need not be included in the operation. Allowing the execution time to be reduced by one clock cycle.

The initial product will also be off by a single bit position.

Example Design

The example state machine provided represents the *Simple Sequential Multiplier*.

- Both operands (multiplier and multiplicand) and product are the same fixed-point format.
- Format is scalable Qm.n, integer and fractional bit lengths as generic parameters, defaulting to Q31.32, with a range to Q63.64.
- Signed two's complement numbers are supported for both the operands and product. All conversions are handled and managed internally by the state machine.
- Rounding can be enabled or disabled, but defaults to enabled; round-to-nearest-even.

As configured for Q31.32, test builds were run using Xilinx ISE. The following performance figures were reached with the corresponding parts.

Xilinx Spartan XC3s500 greater than 100MHz
Xilinx Virtex XC5vlx30 greater than 250MHz

```vhdl
------------------------------------------------
--
-- SimpleSequentialMultiplier.vhd
--
------------------------------------------------
library IEEE;
use IEEE.std_logic_1164.all;
use IEEE.numeric_std.all;

entity SimpleSequentialMultiplier is
--
--   Qmn fixed point format is used.
--
--          sign      binary point
--           |            |
--   Format <s>(integer bits).(fractional bits)
--           _____/ _____/
--              INT_SIZ         FRAC_SIZ
--
--
generic (INT_SIZ: integer range 0 to 63 := 31;
         FRAC_SIZ: integer range 0 to 64 := 32;
         ROUNDING: std_logic := '1');
port
(
    clk: in std_logic; -- system clock
    rst: in std_logic;  -- system reset (must be synchronous)
    -- inputs
    start: in std_logic; -- start multiplication
    multiplier: in signed((1 + INT_SIZ + FRAC_SIZ)-1 downto 0);
    multiplicand: in signed((1 + INT_SIZ + FRAC_SIZ)-1 downto 0);
    -- ouputs
    cmplt: out std_logic; -- multiplication complete
    ovrflw: out std_logic; -- overflow error
    product: out signed((1 + INT_SIZ + FRAC_SIZ)-1 downto 0)
);
end SimpleSequentialMultiplier;

architecture RTL of SimpleSequentialMultiplier is

------------------------
--  Declared constants
------------------------
constant D_SIZ: integer := 1+INT_SIZ+FRAC_SIZ; -- data word size
```

```vhdl
constant A_SIZ: integer := 1+(2*INT_SIZ)+(2*FRAC_SIZ); -- accumulator size

constant D_MSB: integer := D_SIZ-1; -- data word msb bit position
constant A_MSB: integer := A_SIZ-1; -- accumulator msb bit position

constant D_OVR: integer := D_SIZ; -- data word overflow bit position

constant P_LSB: integer := FRAC_SIZ; -- product lsb bit position
constant P_MSB: integer := D_MSB+P_LSB; -- product msb bit position


----------------------
-- Declared signals
----------------------
signal w1: signed(D_MSB downto 0) := (others=>'0');
signal w2: signed(D_MSB downto 0) := (others=>'0');
signal acc: signed(A_MSB downto 0) := (others=>'0');
signal grs: unsigned(2 downto 0) := (others=>'0');

signal cnt: integer range 0 to D_SIZ-1 := 0;
signal sign: std_logic := '0';
signal busy: std_logic := '0';


----------------------
-- Enumeration lists
----------------------
type sm_def is
(
    RESET,
    START_MUL,
    MUL,
    MUL2,
    MUL3,
    ROUND,
    ROUND2
);
signal state: sm_def := RESET;

-------------------------------- module code ----------------------------
begin


-------------------------------------------------
--
-- Simple Sequential Multiplier state machine
```

```vhdl
--
-------------------------------------------------
process(rst,clk)
begin

    if(rst='1') then

        -- working registers
        w1  <= (others=>'0');
        w2  <= (others=>'0');
        acc <= (others=>'0');
        grs <= (others=>'0');

        -- local signals
        sign <= '0';

        -- handshake signals
        busy <= '0';
        ovrflw <= '0';

        -- states
        state <= RESET;

    elsif rising_edge(clk) then

        --
        --  State machine body
        --
        case state is
            -- reset state
            when RESET =>
                state <= START_MUL;
            --
            -- Multiplier body
            --
            when START_MUL =>
                if(start = '1') then
                    w1 <= multiplier;
                    w2 <= multiplicand;
                    busy <= '1';
                    ovrflw <= '0';
                    cnt <= 0;
                    state <= MUL;
                end if;
```

21

```vhdl
-- multiplication operation (w1->multiplier, w2->multiplicand)
when MUL =>
    -- clear accumulator
    acc <= (others=>'0');
    grs <= (others=>'0');
    -- save resultant sign
    sign <= w1(w1'high) xor w2(w2'high);
    -- negate w1 operand
    if(w1(w1'high) = '1') then
        w1 <= (not w1) + 1;
    end if;
    -- negate w2 operand
    if(w2(w2'high) = '1') then
        w2 <= (not w2) + 1;
    end if;
    state <= MUL2;
when MUL2 =>
    -- right-shift multiplier for next cycle
    w1 <= '0'&w1(w1'high downto 1);
    -- add and right-shift upper accumulator in same cycle
    if(w1(0) = '1') then
        acc(A_MSB downto (A_SIZ-D_SIZ)) <=
            '0'&acc(A_MSB downto (A_SIZ-D_SIZ+1)) + w2;
    else
        acc(A_MSB downto (A_SIZ-D_SIZ)) <=
            '0'&acc(A_MSB downto (A_SIZ-D_SIZ+1)) + 0;
    end if;
    -- right shift lower accumulator with lsb of upper half
    acc(A_SIZ-D_SIZ-1 downto 0) <=
        acc(A_SIZ-D_SIZ downto 1);
    -- right shift guard, round, and sticky bits
    grs <= acc(P_LSB)&grs(2)&(grs(1) or grs(0));

    -- sequence counter
    if(cnt < D_SIZ-1) then
        cnt <= cnt + 1;
    else
        state <= MUL3;
    end if;
when  MUL3 =>
    -- normalize result back into accumulator
    acc(D_OVR downto 0) <= '0'&acc(P_MSB downto P_LSB);

    -- set overflow flag
```

```vhdl
                if(acc(A_MSB downto P_MSB) /= 0) then
                    ovrflw <= '1';
                    busy <= '0';
                    state <= START_MUL;
                else
                    state <= ROUND;
                end if;
            -- rounding for multiply
            when ROUND =>
            -- round result to nearest even number
            if (ROUNDING = '1' and(grs > 4 or (grs = 4 and acc(0) = '1'))) then
                    acc(D_OVR downto 0) <= '0'&acc(D_MSB downto 0) + 1;
                end if;
                state <= ROUND2;
            when ROUND2 =>
                if(acc(D_OVR downto D_MSB) /= 0) then
                    ovrflw <= '1';
                elsif(sign = '1') then
                    acc(D_MSB downto 0) <= (not acc(D_MSB downto 0)) + 1;
                end if;
                busy <= '0';
                state <= START_MUL;

            when others =>
                state <= RESET;
        end case;

    end if;

end process;

-- output signals
cmplt <= (not start) and (not busy);
product <= acc(D_MSB downto 0);

end RTL;
```

6 Booth Sequential Multiplier

The *Booth Sequential Multiplier* algorithm is very similar to the *Simple Sequential Multiplier* covered in the Section 5. Operating on signed two's complement numbers directly is its primary advantage. No sign conversions on operands or product are required. This Results in fewer overall clock cycles because there is no conversion overhead, and makes it a more efficient design in terms of logic resource use.

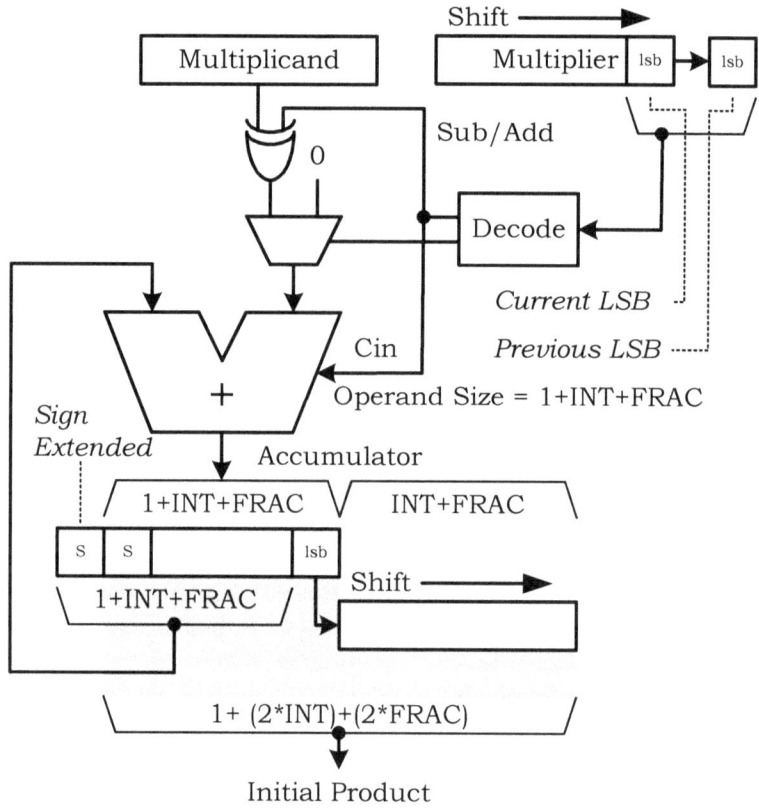

Figure 7

Figure 7 illustrates multiplying two signed fixed-point numbers. The Booth algorithm requires an additional bit to store the multiplier LSB of the previous cycle. On each cycle, the LSB of the multiplier along with the stored LSB from the previous cycle are evaluated then right-shifted. The extra bit is updated with the current LSB of the multiplier while the next bit in the multiplier register is brought into the LSB position.

These two bits determine the type of operation, as shown in the table below. The multiplicand is either added or subtracted, or zero is added, to the upper half of the accumulator (offset by 1 and sign extended). Constantly extending the sign bit is another requirement of the Booth algorithm.

Current	Previous	Operation
0	0	Add zero
0	1	Add multiplicand
1	0	Subtract multiplicand
1	1	Add zero

The sum or difference updates the upper half of the accumulator while the LSB of the previous result is right-shifted into the lower half of the accumulator as the entire lower portion is right-shifted.

Clock cycles required to complete the operation is equal to the number of bits in the multiplier, LSB through MSB, resulting in the initial product being present in the combined upper and lower accumulator.

Note: *see chapter 4 for Normalizing, Rounding, and Overflow.*

Because round-to-nearest-even applies equally to both signed-magnitude and two's complement numbers, it can be used here as well.

Optional Improvements

Expanding the design of the *Booth Sequential Multiplier* to be more like the *Combined Partial Product Multiplier* in chapter 7, would cut the number of clock cycles by half.

This it would require a custom adder very similar to the *carry save adder* (CSA) but with two carry inputs, thus allowing individual control of two's complement math on both partial products.

A more sophisticated approach would be to implement a Booth's recoding scheme that supports a Radix-4 or bit-pairing operations or higher. Published information is available to do so.

Example Design

The example state machine provided represents the *Booth Sequential Multiplier*.

- Both operands (multiplier and multiplicand) and product have the same fixed-point format.
- Format is scalable Qm.n, integer and fractional bit lengths as generic parameters, defaulting to Q31.32, with a range to Q63.64.
- Signed two's complement numbers are supported for both the operands and product. No conversion is required.
- Rounding can be enabled or disabled, but defaults to enabled; round-to-nearest-even is used.

Performance is virtually identical to the *Simple Sequential Multiplier* because the carry propagation distance is the same. The additional decode and subtraction function may create some extra delay, but it is usually negligible.

```vhdl
----------------------------------------------------
--
--    BoothSequentialMultiplier.vhd
--
----------------------------------------------------
library IEEE;
use IEEE.std_logic_1164.all;
use IEEE.numeric_std.all;

entity BoothSequentialMultiplier is
--
--    Qmn fixed point format is used.
--
--          sign      binary point
--           |            |
--    Format <s>(integer bits).(fractional bits)
--           _____/ _____/
--              INT_SIZ          FRAC_SIZ
--
--
generic (INT_SIZ: integer range 0 to 63 := 31;
         FRAC_SIZ: integer range 0 to 64 := 32;
         ROUNDING: std_logic := '1');
port
(
    clk: in std_logic; -- system clock
    rst: in std_logic; -- system reset (must be synchronous)
    -- inputs
    start: in std_logic; -- start multiplication
    multiplier: in signed((1 + INT_SIZ + FRAC_SIZ)-1 downto 0);
    multiplicand: in signed((1 + INT_SIZ + FRAC_SIZ)-1 downto 0);
    -- ouputs
    cmplt: out std_logic; -- multiplication complete
    ovrflw: out std_logic; -- overflow error
    product: out signed((1 + INT_SIZ + FRAC_SIZ)-1 downto 0)
);
end BoothSequentialMultiplier;

architecture RTL of BoothSequentialMultiplier is


------------------------
--   Declared constants
------------------------
constant D_SIZ: integer := 1+INT_SIZ+FRAC_SIZ; -- data word size
```

```vhdl
constant A_SIZ: integer := 1+(2*INT_SIZ)+(2*FRAC_SIZ); -- accumulator size

constant D_MSB: integer := D_SIZ-1; -- data word msb bit position
constant A_MSB: integer := A_SIZ-1; -- accumulator msb bit position

constant D_OVR: integer := D_SIZ; -- data word overflow bit position

constant P_LSB: integer := FRAC_SIZ; -- product lsb bit position
constant P_MSB: integer := D_MSB+P_LSB; -- product msb bit position

----------------------
-- Declared signals
----------------------
signal w1: signed(D_MSB downto 0) := (others=>'0');
signal w2: signed(D_MSB downto 0) := (others=>'0');
signal acc: signed(A_MSB downto 0) := (others=>'0');
signal grs: unsigned(2 downto 0) := (others=>'0');

signal cnt: integer range 0 to D_SIZ-1 := 0;
signal mlsb: std_logic := '0';
signal busy: std_logic := '0';


----------------------
-- Enumeration lists
----------------------
type sm_def is
(
    RESET,
    START_MUL,
    MUL,
    MUL2,
    ROUND,
    ROUND2
);
signal state: sm_def := RESET;

----------------------------------- module code ----------------------------
begin

-------------------------------------------------
--
-- Booth Sequential Multiplier state machine
--
-------------------------------------------------
```

```vhdl
process(rst,clk)
begin

    if(rst='1') then

        -- working registers
        w1  <= (others=>'0');
        w2  <= (others=>'0');
        acc <= (others=>'0');
        grs <= (others=>'0');

        -- local signals
        mlsb <= '0';

        -- handshake signals
        busy <= '0';
        ovrflw <= '0';

        -- states
        state <= RESET;

    elsif rising_edge(clk) then

        --
        --   State machine body
        --
        case state is
            -- reset state
            when RESET =>
                state <= START_MUL;
            --
            -- Multiplier body
            --
            when START_MUL =>
                if(start = '1') then
                    w1 <= multiplier;
                    w2 <= multiplicand;
                    mlsb <= '0';
                    acc <= (others=>'0');
                    grs <= (others=>'0');
                    busy <= '1';
                    ovrflw <= '0';
                    cnt <= 0;
                    state <= MUL;
```

```vhdl
        end if;
-- multiplication operation (w1->multiplier, w2->multiplicand)
when MUL =>
    -- right-shift multiplier for next cycle
    w1 <= '0'&w1(w1'high downto 1);
    mlsb <= w1(0);
    -- add/sub and right-shift upper accumulator in same cycle
    if(w1(0) = '0' and mlsb = '1') then -- add multiplicand
        acc(A_MSB downto (A_SIZ-D_SIZ)) <=
            acc(A_MSB)&acc(A_MSB downto (A_SIZ-D_SIZ+1)) + w2;
    elsif(w1(0) = '1' and mlsb = '0') then -- subtract multiplicand
        acc(A_MSB downto (A_SIZ-D_SIZ)) <=
            acc(A_MSB)&acc(A_MSB downto (A_SIZ-D_SIZ+1)) - w2;
    else -- add zero
        acc(A_MSB downto (A_SIZ-D_SIZ)) <=
            acc(A_MSB)&acc(A_MSB downto (A_SIZ-D_SIZ+1)) + 0;
    end if;
    -- right shift lower accumulator with lsb of upper half
    acc(A_SIZ-D_SIZ-1 downto 0) <=
        acc(A_SIZ-D_SIZ downto 1);

    -- right shift guard, round, and sticky bits
    -- (based on product sign)
    if(cnt > 0) then
        grs(2 downto 1) <= acc(P_LSB)&grs(2);
        if(sign = '0') then
            grs(0) <= grs(1) or grs(0);-- (+) product
        else
            grs(0) <= not(not(grs(1)) or not(grs(0)));-- (-) product
        end if;
    end if;

    -- sequence counter
    if(cnt < D_SIZ-1) then
        cnt <= cnt + 1;
    else
        state <= MUL2;
    end if;
when  MUL2 =>
    -- normalize result back into accumulator
    acc(D_OVR downto 0) <=
        acc(P_MSB)&acc(P_MSB downto P_LSB);
    state <= ROUND;
    -- set overflow flag based in sign
```

```vhdl
                    for i in A_MSB downto P_MSB loop
                        if(acc(i) /= acc(P_MSB)) then
                            ovrflw <= '1';
                            busy <= '0';
                            state <= START_MUL;
                        end if;
                    end loop;
                -- rounding for multiply
                when ROUND =>
                    if(ROUNDING = '1') then
                        -- round to nearest even regardless of sign
                        if(grs > 4 or (grs = 4 and acc(0) = '1')) then
                            acc(D_OVR downto 0) <= '0'&acc(D_MSB downto 0) + 1;
                        end if;
                    end if;
                    state <= ROUND2;
                when ROUND2 =>
                    -- check for overflow from rounding
                    if(acc(D_OVR) /= acc(D_MSB)) then
                        ovrflw <= '1';
                    -- check for negative zero
                    elsif(acc(D_MSB) = '1' and acc(D_MSB-1 downto 0) = 0) then
                        ovrflw <= '1';
                    end if;
                    busy <= '0';
                    state <= START_MUL;

                when others =>
                    state <= RESET;
            end case;

        end if;

    end process;

    -- output signals
    cmplt <= (not start) and (not busy);
    product <= acc(D_MSB downto 0);

end RTL;
```

7 Combined Partial Product Multiplier

The *Simple Sequential Multiplier* in section 5 operates by adding individual partial products on every clock cycle, as pointed out on the left side of Figure 8; whereas a *Combined Partial Product Multiplier* operates by adding two at a time, as pointed out on the right side in Figure 8. Combining two partial products effectively reduces the number of clock cycles required to complete the multiply by half.

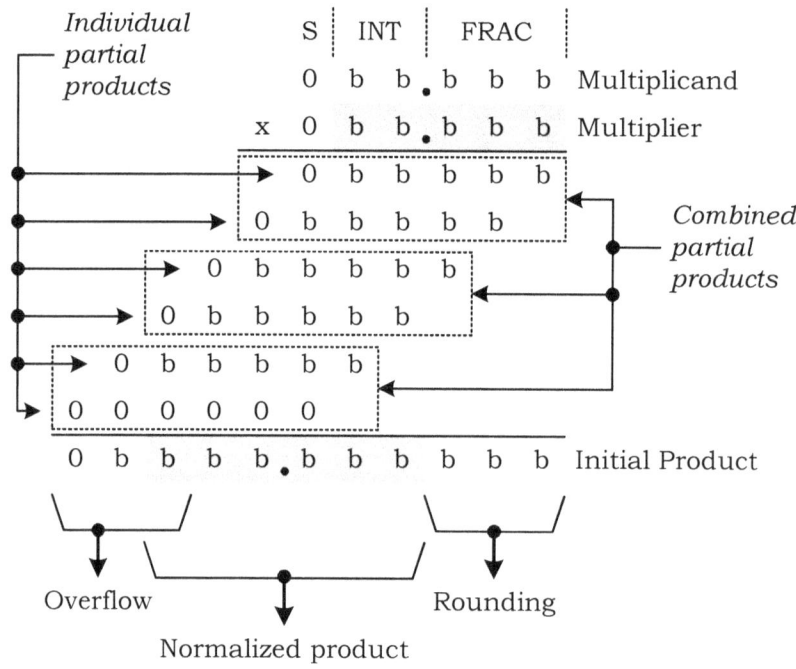

Figure 8

In order to accomplish this, the adder must accept three inputs: two individual partial products and the current value of the accumulator. Conjoining a *Carry Save Adder* (CSA) and a standard or traditional adder (ripple carry), creates a three input adder, as shown in Figure 9.

A single stage CSA accepts and adds the 3 inputs with the resulting sums and carries propagated down. The standard adder combines the sum and carries to produce a final product representing all three operands. This idea is easily extended to multiple stages of CSA blocks -- as is the case in many advanced floating-point execution units.

While the extra level of logic from the CSA introduces a slight delay, the maximum clock speed should not be significantly effected.

Note: *In Figure 9 the carry out from both the CSA and standard adder are not used, because the implementation in Figure 10 will never overflow.*

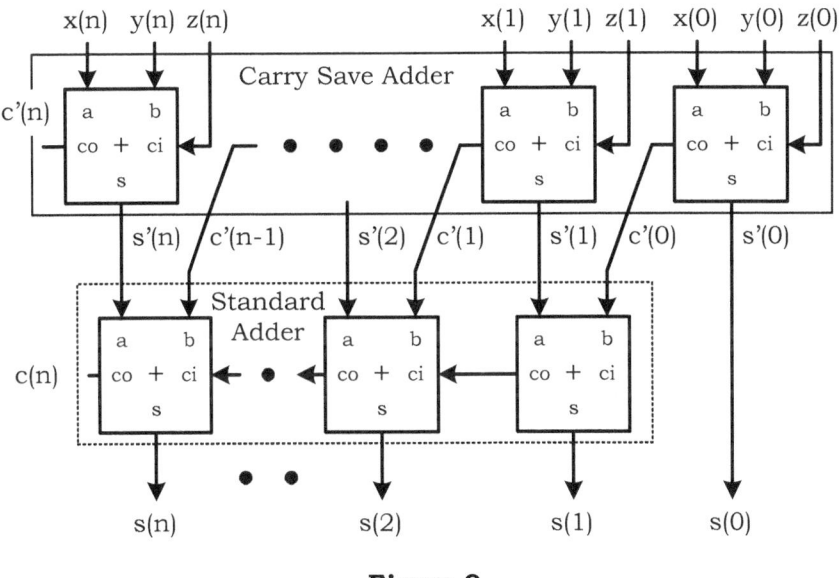

Figure 9

Figure 10 illustrates the implementation of the *Combined Partial Product Multiplier*.

On every clock cycle, the lower two bits of the multiplier are evaluated, LSB and LSB+1, then is right-shifted by two for the next cycle. If the corresponding bits are '1', the multiplicand is gated into the CSA, otherwise a zero is gated in. Notice the LSB gated multiplicand is padded by a leading zero; whereas the LSB+1 gated multiplicand is offset by a zero. This allows multiplicand data to align with the double shifts of the accumulator.

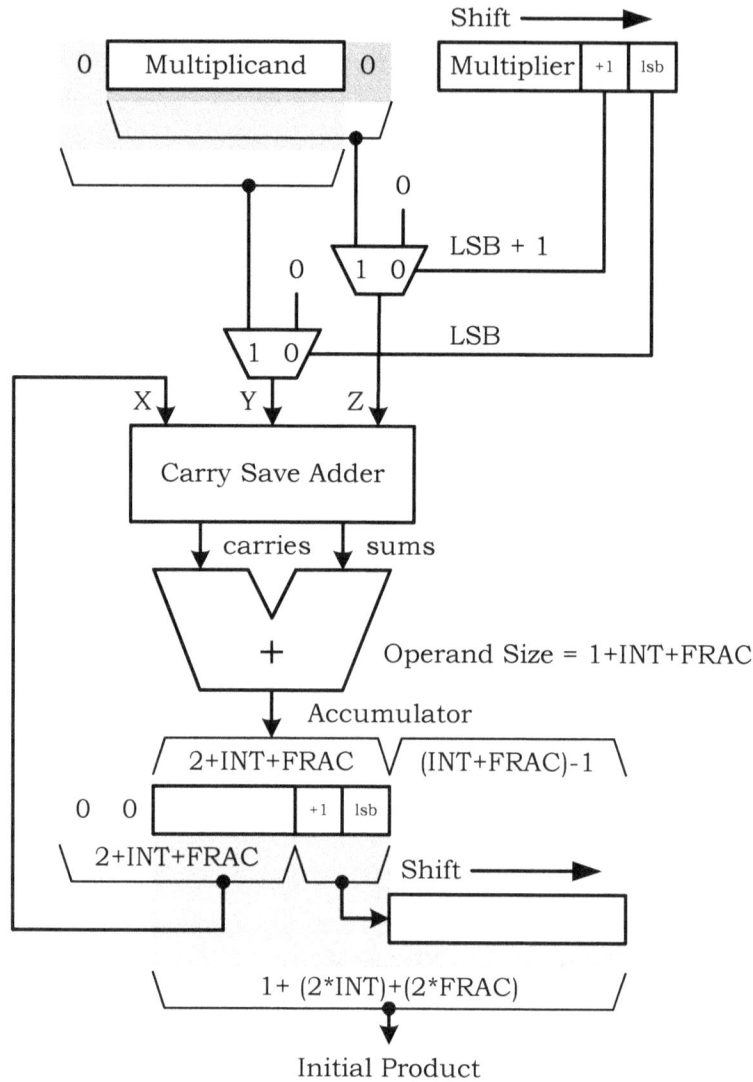

Figure 10

At the same time, the value in the upper portion of the accumulator, padded by two leading zeros, is fed back into the CSA, while its lower two bits are right-shifted into the lower half of the accumulator.

Partial products are added and shifted two at a time, except when the operand size has an odd number of bits. In this instance, when the last bit in the multiplier is being evaluated, the fed back portion of the accumulator is only offset by a single leading zero, and only the LSB of the upper portion of the accumulator is right shifted into the lower half. This exception is not included in Figure 10.

The number of clock cycles required to complete the operation is equal to the number of bits in the multiplier divided by two, resulting in the initial product being present in the combined upper and lower accumulator.

Note: *see chapter 4 for Normalizing, Rounding, and Overflow.*

Optional Improvements

Extend the number of partial products beyond two. This typically involves CSA trees of two or more levels, along with expanding the traditional adder length.

Exclude the sign bit in the operation. This is only effective if operand size has an odd number of bits.

Example Design

The example state machine provided represents the *Combined Partial Product Multiplier*.

- Both operands (multiplier and multiplicand) and product are the same fixed-point format.
- Format is scalable Qm.n, integer and fractional bit lengths as generic parameters, defaulting to Q31.32.
- Signed two's complement numbers are supported for both the operands and product. All conversions are handled and managed internally by the state machine.
- Rounding can be enabled or disabled, but defaults to enabled; round-to-nearest-even.

As configured Q31.32, test builds were run using Xilinx ISE. The following performance figures were reached with the corresponding parts.

Xilinx Spartan XC3s500 greater than 100MHz
Xilinx Virtex XC5vlx30 greater than 245MHz

```vhdl
----------------------------------------------------
--
--  CombinedPartialProductMultiplier.vhd
--
----------------------------------------------------
library IEEE;
use IEEE.std_logic_1164.all;
use IEEE.numeric_std.all;

entity CombinedPartialProductMultiplier is
--
--   Qmn fixed point format is used.
--
--          sign      binary point
--           |            |
--   Format <s>(integer bits).(fractional bits)
--           _____/ _____/
--              INT_SIZ        FRAC_SIZ
--
--
generic (INT_SIZ: integer range 0 to 63 := 31;
         FRAC_SIZ: integer range 0 to 64 := 32;
         ROUNDING: std_logic := '1');
port
(
    clk: in std_logic; -- system clock
    rst: in std_logic;  -- system reset (must be synchronous)
    -- inputs
    start: in std_logic; -- start multiplication
    multiplier: in signed((1 + INT_SIZ + FRAC_SIZ)-1 downto 0);
    multiplicand: in signed((1 + INT_SIZ + FRAC_SIZ)-1 downto 0);
    -- ouputs
    cmplt: out std_logic; -- multiplication complete
    ovrflw: out std_logic; -- overflow error
    product: out signed((1 + INT_SIZ + FRAC_SIZ)-1 downto 0)
);
end CombinedPartialProductMultiplier;

architecture RTL of CombinedPartialProductMultiplier is
------------------------
--  Declared constants
------------------------
constant D_SIZ: integer := 1+INT_SIZ+FRAC_SIZ; -- data word size
constant A_SIZ: integer := 1+(2*INT_SIZ)+(2*FRAC_SIZ); -- accumulator size
```

```vhdl
constant D_MSB: integer := D_SIZ-1; -- data word msb bit position
constant A_MSB: integer := A_SIZ-1; -- accumulator msb bit position

constant D_OVR: integer := D_SIZ; -- data word overflow bit position

constant P_LSB: integer := FRAC_SIZ; -- product lsb bit position
constant P_MSB: integer := D_MSB+P_LSB; -- product msb bit position


----------------------
-- Declared signals
----------------------
signal w1: signed(D_MSB downto 0) := (others=>'0');
signal w2: signed(D_MSB downto 0) := (others=>'0');
signal acc: signed(A_MSB downto 0) := (others=>'0');
signal grs: unsigned(2 downto 0) := (others=>'0');

signal y: unsigned(D_MSB+1 downto 0) := (others=>'0');
signal z: unsigned(D_MSB+1 downto 0) := (others=>'0');
signal x: unsigned(D_MSB+1 downto 0) := (others=>'0');
signal sout: unsigned(D_MSB+1 downto 0) := (others=>'0');
signal cout: unsigned(D_MSB+1 downto 0) := (others=>'0');

signal cnt: integer range 0 to D_SIZ := 0;
signal sign: std_logic := '0';
signal busy: std_logic := '0';


----------------------
-- Enumeration lists
----------------------
type sm_def is
(
    RESET,
    START_MUL,
    MUL,
    MUL2,
    MUL3,
    ROUND,
    ROUND2
);
signal state: sm_def := RESET;

------------------------------------ module code ------------------------------
begin
```

```vhdl
-------------------------------------------------------
--
--   Combined Partial Product Multiplier state machine
--
-------------------------------------------------------
process(rst,clk)
begin

    if(rst='1') then
        -- working registers
        w1  <= (others=>'0');
        w2  <= (others=>'0');
        acc <= (others=>'0');
        grs <= (others=>'0');

        -- local signals
        sign <= '0';

        -- handshake signals
        busy <= '0';
        ovrflw <= '0';

        -- states
        state <= RESET;

    elsif rising_edge(clk) then
        --
        --   State machine body
        --
        case state is
            -- reset state
            when RESET =>
                state <= START_MUL;
            --
            --   Multiplier body
            --
            when START_MUL =>
                if(start = '1') then
                    w1 <= multiplier;
                    w2 <= multiplicand;
                    busy <= '1';
                    ovrflw <= '0';
                    cnt <= D_SIZ; -- start with max count
                    state <= MUL;
```

```vhdl
      end if;
-- multiplication operation (w1->multiplier, w2->multiplicand)
when MUL =>
    -- clear accumulator
    acc <= (others=>'0');
    grs <= (others=>'0');
    -- save resultant sign
    sign <= w1(w1'high) xor w2(w2'high);
    -- negate w1 operand
    if(w1(w1'high) = '1') then
        w1 <= (not w1) + 1;
    end if;
    -- negate w2 operand
    if(w2(w2'high) = '1') then
        w2 <= (not w2) + 1;
    end if;
    state <= MUL2;
when MUL2 =>
    -- update upper half of accumulator with sum and carries
    -- from CSA below (offset carry by 1-bit)
    acc(A_MSB downto (A_SIZ-D_SIZ-1)) <=
        signed(sout) + signed(cout(cout'high-1 downto 0)&'0');
    w1 <= "00"&w1(w1'high downto 2); -- right-shift multiplier

    --  manage accumulator and rounding bits
    if(cnt = 1) then -- last bit of an odd numbered operand
        acc(A_SIZ-D_SIZ-2 downto 0) <=
            acc(A_SIZ-D_SIZ-1 downto 1);
        grs <= acc(P_LSB)&grs(2)&(grs(1) or grs(0));
    else -- normal sequence, shift 2-bits
        acc(A_SIZ-D_SIZ-2 downto 0) <=
            acc(A_SIZ-D_SIZ downto 2);
        grs <=
            acc(P_LSB+1)&acc(P_LSB)&
            (grs(2) or grs(1) or grs(0));
    end if;
    -- manage bit counter
    if(cnt <= 2) then
        -- exit on last shift
        state <= MUL3;
    else
        cnt <= cnt - 2;
    end if;
```

```vhdl
        when MUL3 =>
            -- normalize result back into accumulator
            acc(D_OVR downto 0) <= '0'&acc(P_MSB downto P_LSB);

            -- set overflow flag
            if(acc(A_MSB downto P_MSB) /= 0) then
                ovrflw <= '1';
                busy <= '0';
                state <= START_MUL;
            else
                state <= ROUND;
            end if;
        -- rounding for multiply
        when ROUND =>
        -- round result to nearest even number
        if (ROUNDING = '1' and(grs > 4 or (grs = 4 and acc(0) = '1'))) then
            acc(D_OVR downto 0) <= '0'&acc(D_MSB downto 0) + 1;
            end if;
            state <= ROUND2;
        when ROUND2 =>
            if(acc(D_OVR downto D_MSB) /= 0) then
                ovrflw <= '1';
            elsif(sign = '1') then
                acc(D_MSB downto 0) <= (not acc(D_MSB downto 0)) + 1;
            end if;
            busy <= '0';
            state <= START_MUL;

        when others =>
            state <= RESET;
    end case;

  end if;

end process;

-- output signals
cmplt <= (not start) and (not busy);
product <= acc(D_MSB downto 0);


------------------------------------
--
--   Carry save adder (CSA) circuits
```

```vhdl
--
-------------------------------------
--  carry save adder inputs

-- accumulator output offset by either 1 (remaining odd bit) or 2 bits (all others)
x <= unsigned('0'&acc(A_MSB downto (A_SIZ-D_SIZ))) when cnt = 1 else
     unsigned("00"&acc(A_MSB downto (A_SIZ-D_SIZ+1)));

-- multiplicand partial product with leading zero
y <= unsigned('0'&w2) when w1(0) = '1' else (others=>'0');

-- multiplicand partial product offset by zero
z <= unsigned(w2&'0') when w1(1) = '1' else (others=>'0');

--  carry save adder
gen_csa: for i in 0 to D_MSB+1 generate
begin
    -- compute sum out
    sout(i) <= (y(i) xor z(i)) xor x(i);
    -- compute carry out
    cout(i) <= (y(i) and x(i)) or (z(i) and x(i)) or (y(i) and z(i));

end generate;


end RTL;
```

8 Combined Array Multiplier

The highest performance multipliers are combinational array multipliers, which are typically composed of an array of full-adders – outputting results in a single clock cycle. There are slight variations between types.

Implementing an array with general gate resources is appropriate for working with small numbers, but large operands will require an enormous number of gates and would be ineffectively slow. Many FPGA vendors, as well as ASIC foundries, offer embedded or hard IP array multipliers that are specifically designed for speed and do not consume general gate or routing resources.

Configurations vary, but a common configuration for an embedded array multiplier is signed 18-bit x 18-bit with a 36-bit product. The problem is obvious: an 18-bit operand is typically too small for fixed or floating-point designs. Nonetheless, by properly combining multiple arrays, larger operands can be supported.

Combining arrays involves dividing operands, sometimes referred to as splitting, into slices small enough fit into the inputs of individual arrays, then recombining the results, or sub-products of each output through adders.

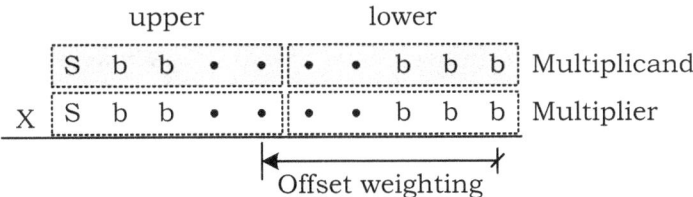

Figure 11

Figure 11 illustrates this concept. Both the multiplicand and multiplier are split into two parts, an upper and lower slice. Offset weighting means the power-of-two that the LSB of that slice has must be maintained when recombining sub-products during the adding phase.

As shown in Figure 12, each slice combination is multiplied first, then the resulting sub-products are added together.

Sub-product 1 has no offset and is treated as a positive number. Both sub-products 2 and 3 are offset with the number of zeros equal to the number of bits in the lower slice (offset weighting) and with the sign extended to the full product size, which is 2x the operand size. Finally, sub-product 4 is offset by 2x the number of zeros in the lower slice. Sign extension is not needed because the sign bit already occupies the sub-product MSB position. Subsequently, all are added together. Although counter-intuitive, all non-sign bits in each slice retain their bit weight, regardless of the operand's sign.

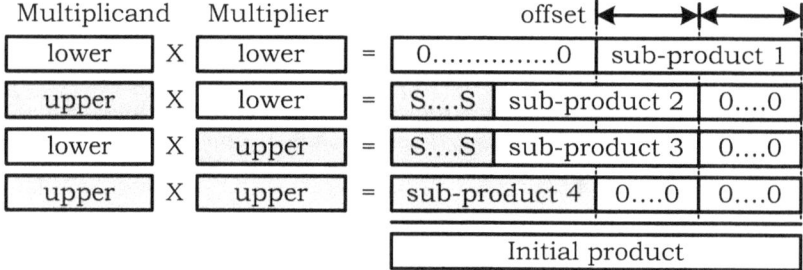

Figure 12

Notice, that each time the multiplication involves an upper slice the sign is preserved or extended. All other operations treat the sub-product as a positive number. While this example has only one positive sub-product, designs where the operands are split into more than two slices will have many.

Even with the speed advantage of using array multipliers, adding the resulting sub-products still involve standard adders, and must be pipelined to break-up the accumulative ripple carry propagation from top to bottom, as diagramed in Figure 13.

Each of the four multiplications are done concurrently with four separate multipliers, registering each sub-product (except for the lowest one). These sub-products are then offset and sign extended as needed then passed through a multistage adder. The limit of this design is an operand ranging from 19 to 36-bits, but it will execute in only three clocks and requires no state machine.

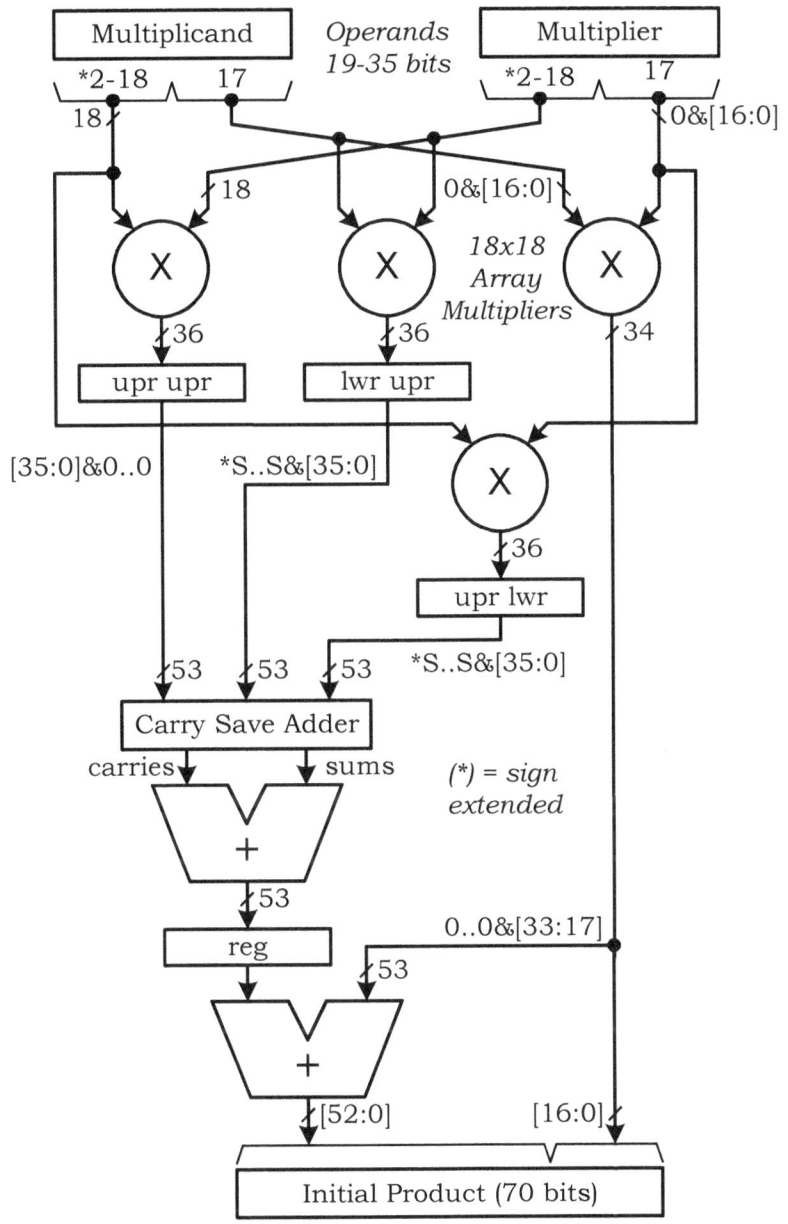

Figure 13

Note: *Because the arrays are signed 18 x18-bit multipliers, sub-products requiring positive numbers must set the MSB of each input to zero, which translates into a 17-bit x 17-bit* multiplication. *The 36-bit sub-product is reduced to 34-bits by discarding the upper two zeros.*

Figure 14 more clearly depicts how each sub-product is properly aligned before addition. Sub-product 1 is a 34-bit product, bits [16:0] pass directly through becoming the lower portion of the initial product; whereas bits [33:17] are added as a positive number to the other sub-products. The remaining sub-products 2, 3, and 4 are 36-bit values but are treated as 53-bit (36-bit sub-product + 17 bits); sub-products 2 and 3 are signed extended to 53-bits, and sub-product 4 is instead offset by 17 zeros.

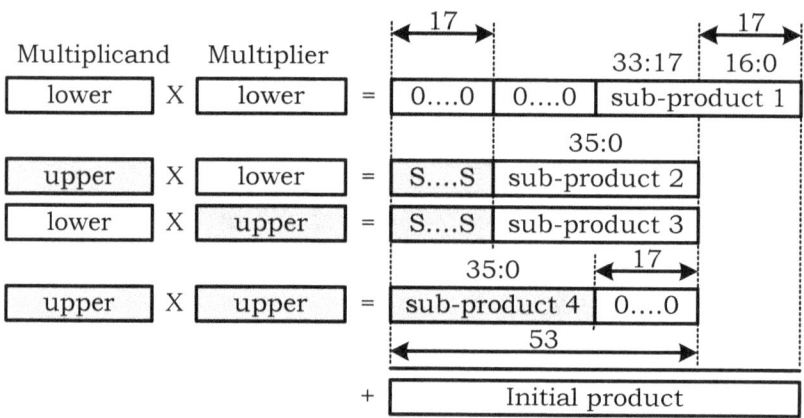

Figure 14

Figure 15 represents a more balanced design between resource use and speed. Pipelining allows different operand sizes to use the same logic mechanism to execute the multiplication.

Depending on the operand size, 19 to 69-bits, the number of slices are auto-configured to 2, 3, or 4. All lower slices are 17-bit with the upper slices internally sign extended to 18-bits. Two sets of multiplexors select and register slice combinations, one from the multiplicand and one from the multiplier. These supply the inputs of two separate array multipliers. This allows two sub-products to be generated in a single clock, which are also registered.

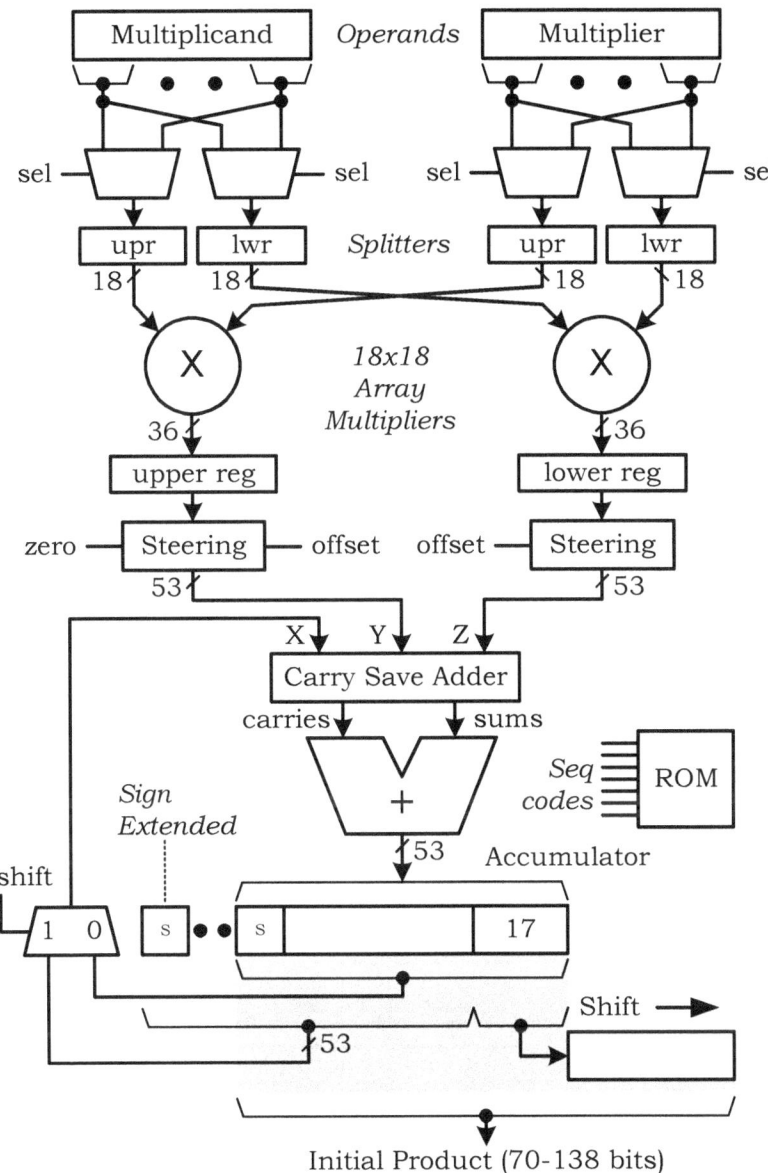

Figure 15

Since sub-products are paired, steering logic allows either one or both sub-product to be individually offset by 17 zeros. In configurations with an odd number of sub-products (slices), the last sub-product is paired with zero.

The adjusted sub-product pair is then added to the accumulator. Each time the accumulator is updated, its current value may or may not be right shifted 17-bits, emulating a paired offset of 17-bits (as opposed to the previous individual offset).

Figures 16, 17, and 18 represent three different configurations of operand sizes, and the number and size of slices used to create sub-products. Figures 19, 20, and 21 illustrate the progression of sub-products through the pipeline, how they are paired, and offsets within pairs and in relationship to other pairs. Offsets can be applied through the steering logic, or by means of shifting the accumulator.

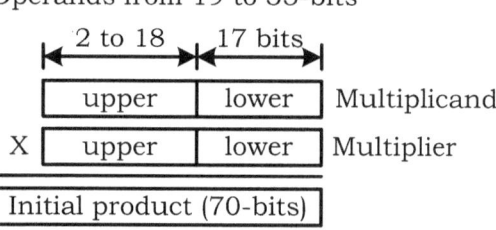

Figure 16

Figure 16 supports operands from 19 to 35 bits, which are split into a lower half of 17-bits and an upper half of 2 to 18-bits. The upper slice is always internally sign extended to 18-bits.

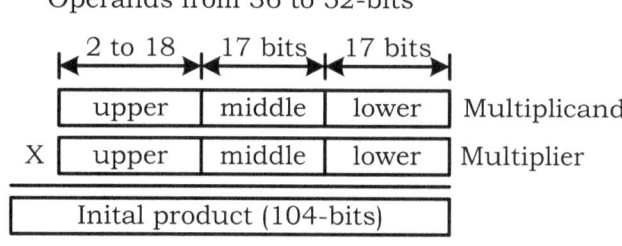

Figure 17

Figure 17 supports operands from 36 to 52-bits. Two lower slices of 17-bits each and an upper slice of 2 to 18-bits.

Operands from 53 to 69-bits

|← 2 to18 →|← 17 bits →|← 17 bits →|← 17 bits →|

| upper | up mid | low mid | lower | Multiplicand
X | upper | up mid | low mid | lower | Multiplier

Initial product (138-bits)

Figure 18

Figure 18 supports operands from 53 to 69-bits, three lower slices of 17-bits each, and an upper slice from 2 to 18-bits.

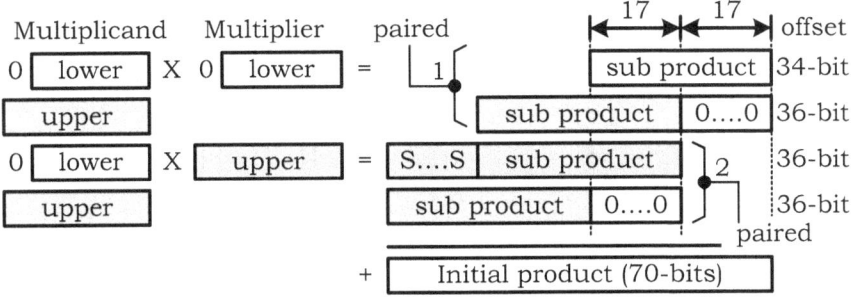

Figure 19

Figure 19 illustrates the progression of the splitting represented in Figure 16. The 53-bit sum pair 1 first initializes the accumulator. On the next clock the 53-bit sum of pair 2 is added to the upper portion of the accumulator right-shifted by 17-bits. The resulting sum updates the upper portion of the accumulator while the lower accumulator is right-shift by 17 bits.

Similarly, Figure 20 shows the progression represented in Figure 17, but has five adds instead of two. First, the 53-bit sum of pair 1 initializes the accumulator. The sums of pairs 2 through 4 are added to the value in the upper accumulator right shifted by 17-bits. The resulting sum updates the upper portion of the accumulator while the lower accumulator is right-shifted. Finally, pair 5, which is paired with zero, is added to the upper portion of the accumulator and not right-shifted.

STATE MACHINES IN VHDL *Multipliers* Vol. 2.1

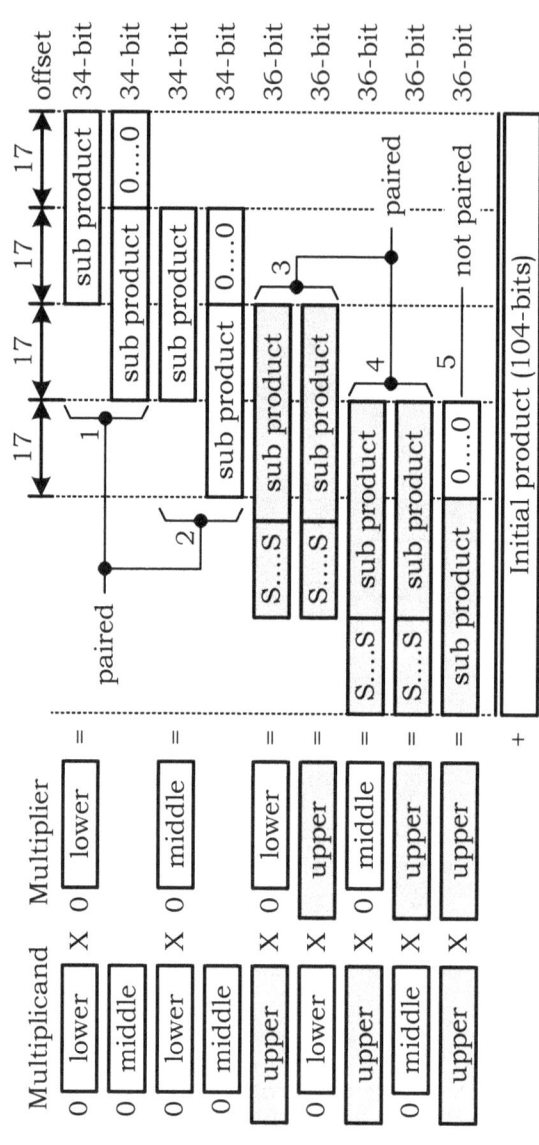

Figure 20

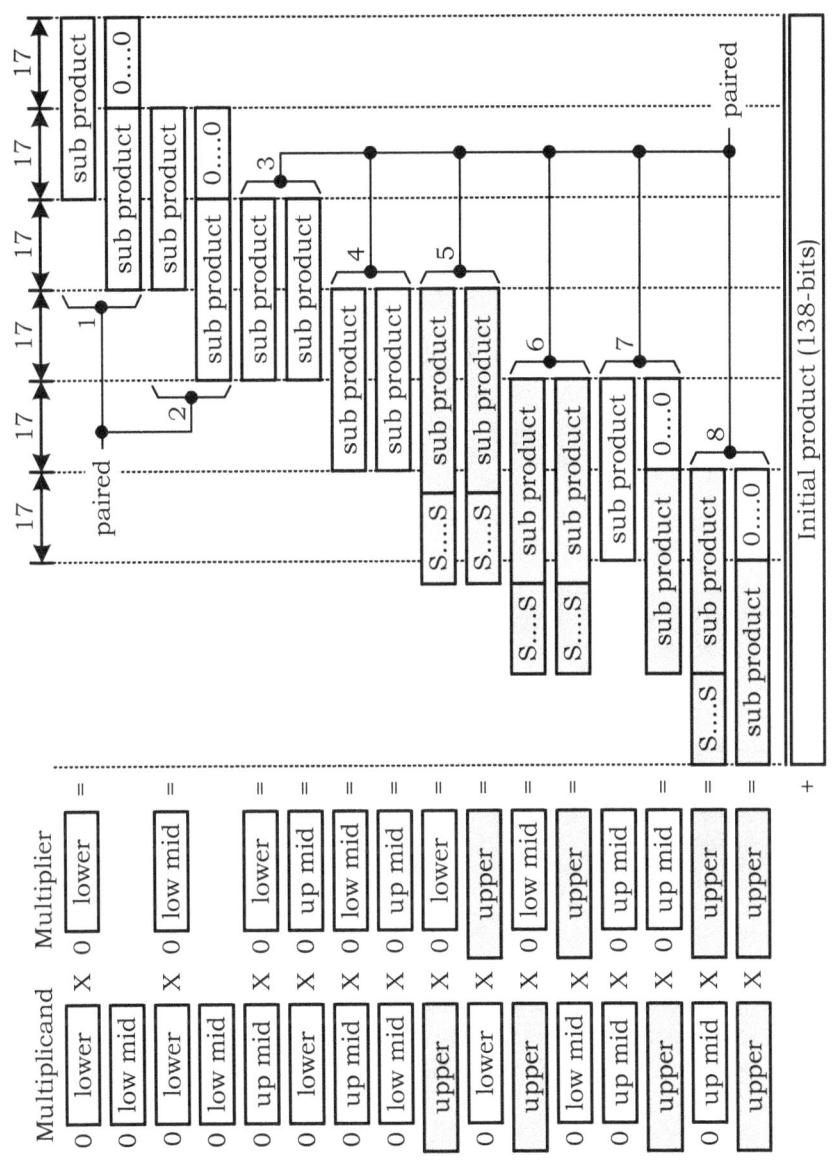

Figure 21

Figure 21 diagrams the progression represented in Figure 18, but has eight adds instead of two or five. First, the 53-bit sum of pair 1 initializes the accumulator. The 53-bit sums of pairs 2, 3, 4, 6, and 8 are added to the value in the upper accumulator right shifted by

17-bits. When updating the upper portion of the accumulator the lower accumulator is right-shifted by 17-bits. Pairs 5 and 7 are added to the upper portion of the accumulator directly without any shifting.

In all configurations, following the execution, the resulting initial product is present in the combined upper and lower accumulator.

> Note: *see chapter 4 for Normalizing, Rounding, and Overflow.*

The number of clock cycles required to complete the operation is 4 clocks for operands between 19 and 35-bits; 7 clocks for 36 to 52-bits; 10 clocks for 53 to 69-bits, plus rounding.

Optional Improvements

Extending the number of embedded multipliers to more than two, thus increasing the number of sub-products that can be processed in one clock. This also requires a multilevel CSA adder or tree.

Another alternative is to use larger embedded multipliers beyond the 18 x 18 configuration, which are becoming more common.

Example Design

The example state machine provided represents the *Combined Array Multiplier.*

- Both operands (multiplier and multiplicand) and product are the same fixed-point format.
- Format is scalable Qm.n, integer and fractional bit lengths as generic parameters, defaulting to Q31.32. The combined integer and fractional portions, plus a sign bit, must be between 19 and 69, which is the operand size.
- Signed two's complement numbers are supported for both the operands and product. No conversion is requited.
- Rounding can be enabled or disabled, but defaults to enabled; round-to-nearest-even is used.
- An added feature is a section of sequence code bits within an inferred ROM, which are indexed by a counter in the state machine. These bits control the multiplexors, steering logic, and shifting, and track with the actions of the state machine.

As configured Q31.32, test builds were run using Xilinx ISE. The following performance figures were reached with the corresponding parts.

Xilinx Spartan XC3s500 greater than 100MHz
Xilinx Virtex XC5vlx30 greater than 240MHz

```vhdl
---------------------------------------------------
--
--   CombinedArrayMultiplier.vhd
--
---------------------------------------------------
library IEEE;
use IEEE.std_logic_1164.all;
use IEEE.numeric_std.all;

entity CombinedArrayMultiplier is
--
--   Qmn fixed point format is used.
--
--          sign      binary point
--            |           |
--   Format <s>(integer bits).(fractional bits)
--            _____/ _____/
--              INT_SIZ        FRAC_SIZ
--
--       Operand Size = 1+INT_SIZ+FRAC_SIZ = range from 19 to 69
--
generic (INT_SIZ: integer range 0 to 68 := 31;
         FRAC_SIZ: integer range 0 to 68 := 32;
         ROUNDING: std_logic := '1');
port
(
    clk: in std_logic; -- system clock
    rst: in std_logic; -- system reset (must be synchronous)
    -- inputs
    start: in std_logic; -- start multiplication
    multiplier: in signed(INT_SIZ+FRAC_SIZ downto 0);
    multiplicand: in signed(INT_SIZ+FRAC_SIZ downto 0);
    -- ouputs
    cmplt: out std_logic; -- multiplication complete
    ovrflw: out std_logic; -- overflow error
    product: out signed(INT_SIZ+FRAC_SIZ downto 0)
);
end CombinedArrayMultiplier;


architecture RTL of CombinedArrayMultiplier is
-----------------------------
--   Basic declared constants
-----------------------------
```

```vhdl
constant OPERAND_SIZ: integer range 19 to 69 := 1+INT_SIZ+FRAC_SIZ;
constant D_SIZ: integer := OPERAND_SIZ; -- data word size
constant D_MSB: integer := D_SIZ-1; -- data word msb bit position
constant D_OVR: integer := D_SIZ; -- data word overflow bit position

constant P_LSB: integer := FRAC_SIZ; -- product lsb bit position
constant P_MSB: integer := D_MSB+P_LSB; -- product msb bit position


--------------------
-- Structure types
--------------------
type sl_array is array (0 to 3) of integer;
type sb_array is array (3 downto 0, 1 downto 0) of integer;
type sc_array is array (0 to 9) of unsigned(12 downto 0);


------------------------------------------
--
-- Computed constants for scaling design
--
------------------------------------------
-- Computes scaling parameters based on operand size
--
-- s(0) = number of 17-bit slices needed
-- s(1) = number of remaining bits in upper slice, 2-18
-- s(2) = maximum sequence count
-- s(3) = size of accumulator
--
function ComputeParameters(operand_size: integer ) return sl_array is
variable s: sl_array := (others=>0);
begin

    if(operand_size >= 19 and operand_size <= 35) then
        s := (1,operand_size-(17*1),3,70);
    elsif(operand_size >= 36 and operand_size <= 52) then
        s := (2,operand_size-(17*2),6,104);
    else -- 53 to 69
        s := (3,operand_size-(17*3),9,138);
    end if;
    return(s);
end function;
--
-- Computes slice boundaries based on number of 17-bit slices
```

```vhdl
--
function ComputeSliceBoundaries(num_17bit_slices: integer;num_upper_bits:
integer) return sb_array is
variable sb: sb_array := (others=>(others=>0));
begin

    if(num_17bit_slices = 1) then
        sb := ((0,0),(0,0),((num_upper_bits-1)+17,17),(16,0));
    elsif(num_17bit_slices = 2) then
        sb := ((0,0),((num_upper_bits-1)+34,34),(33,17),(16,0));
    elsif(num_17bit_slices = 3) then
        sb := (((num_upper_bits-1)+51,51),(50,34),(33,17),(16,0));
    end if;
    return(sb);

end function;
--
--  Computes sequence codes based on number of 17-bit slices
--
--  sc(1:0)  = select for lower muliplier slice
--  sc(3:2)  = select for lower multiplicand slice
--  sc(5:4)  = select for upper muliplier slice
--  sc(7:6)  = select for upper multiplicand slice
--  sc(8)    = lower steering logic offset control
--  sc(9)    = upper steering logic offset control
--  sc(10)   = upper steering logic zero control
--  sc(11)   = enable accumulator
--  sc(12)   = shift accumulator
--
function ComputeSeqCodes(num_17bit_slices: integer) return sc_array is
variable sc: sc_array := (others=>(others=>'0'));
begin
    if(num_17bit_slices = 1) then
        sc := (
            "00000"&"01"&"00"&"00"&"00",
            "00000"&"01"&"01"&"00"&"01",
            "01010"&"00000000",
            "11010"&"00000000",
            "0000000000000",
            "0000000000000",
            "0000000000000",
            "0000000000000",
            "0000000000000",
            "0000000000000");
```

```
    elsif(num_17bit_slices = 2) then
        sc := (
            "00000"&"01"&"00"&"00"&"00",
            "00000"&"01"&"01"&"00"&"01",
            "01010"&"00"&"10"&"10"&"00",
            "11010"&"01"&"10"&"10"&"01",
            "11000"&"00"&"00"&"10"&"10",
            "11000"&"00000000",
            "01101"&"00000000",
            "0000000000000",
            "0000000000000",
            "0000000000000");
    elsif(num_17bit_slices = 3) then
        sc := (
            "00000"&"01"&"00"&"00"&"00",
            "00000"&"01"&"01"&"00"&"01",
            "01010"&"00"&"10"&"10"&"00",
            "11010"&"01"&"10"&"10"&"01",
            "11000"&"00"&"11"&"11"&"00",
            "11000"&"01"&"11"&"11"&"01",
            "01000"&"11"&"10"&"10"&"10",
            "11000"&"11"&"11"&"10"&"11",
            "01010"&"00000000",
            "11010"&"00000000");
    end if;
    return(sc);

end function;
--
-- Update hard constants with parameters from functions
--
constant S: sl_array := ComputeParameters(D_SIZ);
constant N17BITS: integer := S(0); -- number of lower 17-bit slices needed
constant NUPRBITS: integer := S(1); -- number of remaining bits in upper slice
constant MAX_CNT: integer := S(2); -- maximum count needed
constant A_SIZ: integer := S(3); -- accumulator size
constant SB: sb_array := ComputeSliceBoundaries(N17BITS,NUPRBITS);
constant SC: sc_array := ComputeSeqCodes(N17BITS);
constant A_MSB: integer := A_SIZ-1;
```

-- Declared signals

```vhdl
signal mplc: signed(D_MSB downto 0) := (others=>'0');
signal mplr: signed(D_MSB downto 0) := (others=>'0');

signal lwr_mplr_sel: natural range 0 to N17BITS := 0;
signal upr_mplr_sel: natural range 0 to N17BITS := 0;
signal lwr_mplc_sel: natural range 0 to N17BITS := 0;
signal upr_mplc_sel: natural range 0 to N17BITS := 0;

signal lwr_mplr_slice: signed(17 downto 0) := (others=>'0');
signal upr_mplr_slice: signed(17 downto 0) := (others=>'0');
signal lwr_mplc_slice: signed(17 downto 0) := (others=>'0');
signal upr_mplc_slice: signed(17 downto 0) := (others=>'0');

signal upper_reg: signed(35 downto 0) := (others=>'0');
signal lower_reg: signed(35 downto 0) := (others=>'0');

signal lwr_offset: std_logic := '0';
signal upr_zero: std_logic := '0';
signal upr_offset: std_logic := '0';

signal sc_bits: unsigned(12 downto 0):= (others=>'0');

signal x,y,z:signed(52 downto 0) := (others=>'0');
signal sout: std_logic_vector(52 downto 0) := (others=>'0');
signal cout: std_logic_vector(52 downto 0) := (others=>'0');

signal cnt: integer range 0 to SC'high := 0;
signal en_acc: std_logic := '0';
signal shift_acc: std_logic := '0';

signal acc: signed(A_MSB downto 0) := (others=>'0');
signal grs: unsigned(2 downto 0) := (others=>'0');
signal busy: std_logic := '0';


----------------------
-- Enumeration lists
----------------------
type sm_def is
(
    RESET,
    START_MUL,
    PRIME_PIPE,
    MUL,
```

```vhdl
        MULX,
        ROUND,
        ROUND2
);
signal state: sm_def := RESET;



-------------------------------------------------
-- operand slice selector (Splitters)
--
--   Design note: Xilinx XST cannot dynamically
--   determine constant values for slice boundaries;
--   some re-coding required.
-------------------------------------------------
process(rst,clk)
begin

    if(rst='1') then
        upr_mplc_slice <= (others=>'0');
        lwr_mplc_slice <= (others=>'0');
        upr_mplr_slice <= (others=>'0');
        lwr_mplr_slice <= (others=>'0');
    elsif rising_edge(clk) then
        --
        --   multiplier slice mux
        --
        -- lower multiplier slice
        if(lwr_mplr_sel < N17BITS) then
            lwr_mplr_slice <=
                '0'&mplr(SB(lwr_mplr_sel,1) downto SB(lwr_mplr_sel,0));
        else
            lwr_mplr_slice <=
                resize(mplr(SB(N17BITS,1)downto SB(N17BITS,0)),18);
        end if;
        -- upper multiplier slice
        if(upr_mplr_sel < N17BITS) then
            upr_mplr_slice <=
                '0'&mplr(SB(upr_mplr_sel,1) downto SB(upr_mplr_sel,0));
        else
            upr_mplr_slice <=
                resize(mplr(SB(N17BITS,1) downto SB(N17BITS,0)),18);
```

```vhdl
        end if;
        --
        --  multiplicand slice mux
        --
        -- lower muliplicand slice
        if(lwr_mplc_sel < N17BITS) then
            lwr_mplc_slice <=
                '0'&mplc(SB(lwr_mplc_sel,1) downto SB(lwr_mplc_sel,0));
        else -- sign extend
            lwr_mplc_slice <=
                resize(mplc(SB(N17BITS,1) downto SB(N17BITS,0)),18);
        end if;
        -- upper muliplicand slice
        if(upr_mplc_sel < N17BITS) then
            upr_mplc_slice <=
                '0'&mplc(SB(upr_mplc_sel,1) downto SB(upr_mplc_sel,0));
        else -- sign extend
            upr_mplc_slice <=
                resize(mplc(SB(N17BITS,1) downto SB(N17BITS,0)),18);
        end if;
    end if;

end process;


---------------------------------
--  18x18 multipliers (inferred)
---------------------------------
process(rst,clk)
begin

    if(rst='1') then
        upper_reg <= (others=>'0');
        lower_reg <= (others=>'0');
    elsif rising_edge(clk) then
        upper_reg <= upr_mplc_slice * upr_mplr_slice;
        lower_reg <= lwr_mplc_slice * lwr_mplr_slice;
    end if;

end process;


--------------------
--  Steering logic
```

```vhdl
--------------------
-- construct Z data bus
process(lwr_offset,lower_reg)
begin

    -- offset required
    if(lwr_offset = '1') then
        z(16 downto 0) <= (others=>'0'); -- apply offset
        z(52 downto 17) <= lower_reg; -- transfer to upper portion

    else
        z(35 downto 0) <= lower_reg; -- transfer to lower portion
        z(52 downto 36) <= (others=>lower_reg(35)); -- extend sign
    end if;

end process;
-- construct y data bus
process(upr_zero,upr_offset,upper_reg)
begin

    -- zero z data
    if(upr_zero = '1') then
        y <= (others=>'0');
    -- tranfer data to z
    else
        -- offset required
        if(upr_offset = '1') then
            y(16 downto 0) <= (others=>'0'); -- apply offset
            y(52 downto 17) <= upper_reg; -- transfer to upper portion

        else
            y(35 downto 0) <= upper_reg; -- transfer to lower portion
            y(52 downto 36) <= (others=>upper_reg(35)); -- extend sign
        end if;
    end if;

end process;


----------------------
--   Carry save adder
----------------------
-- accumulator output offset by either 17 or none, sign extended
```

```vhdl
x <= resize(acc(A_MSB downto A_MSB-53+17+1),53) when shift_acc = '1' else
     acc(A_MSB downto A_MSB-53+1);

gen_csa: for i in 0 to 52 generate
begin
    -- compute sum out
    sout(i) <= (y(i) xor z(i)) xor x(i);
    -- compute carry out
    cout(i) <= (y(i) and x(i)) or (z(i) and x(i)) or (y(i) and z(i));

end generate;


---------------------------
-- Sequence code signals
---------------------------
lwr_mplr_sel <= to_integer('0'&sc_bits(1 downto 0));
lwr_mplc_sel <= to_integer('0'&sc_bits(3 downto 2));
upr_mplr_sel <= to_integer('0'&sc_bits(5 downto 4));
upr_mplc_sel <= to_integer('0'&sc_bits(7 downto 6));

lwr_offset <= sc_bits(8);
upr_offset <= sc_bits(9);
upr_zero <= sc_bits(10);

en_acc <= sc_bits(11);
shift_acc <= sc_bits(12);


---------------------------------------------
--
-- Combined Array Multiplier state machine
--
---------------------------------------------
process(rst,clk)
begin

    if(rst='1') then

        -- sequence code signals
        sc_bits <= (others=>'0');

        -- working registers
        mplr <= (others=>'0');
```

```vhdl
        mplc <= (others=>'0');
        acc  <= (others=>'0');
        grs  <= (others=>'0');

        -- local signals
        cnt <= 0;

        -- handshake signals
        busy <= '0';
        ovrflw <= '0';

        -- states
        state <= RESET;

    elsif rising_edge(clk) then
        -- maintain muxes at first position
        cnt <= 0;
        --
        --  State machine body
        --
        case state is
            -- reset state
            when RESET =>
                state <= START_MUL;
            --
            --  Multiplier body
            --
            when START_MUL =>
                if(start = '1') then
                    mplr <= multiplier;
                    mplc <= multiplicand;
                    acc <= (others=>'0');
                    grs <= (others=>'0');
                    busy <= '1';
                    ovrflw <= '0';
                    cnt <= cnt + 1;
                    sc_bits <= SC(cnt);
                    state <= PRIME_PIPE;
                end if;
            -- wait for 1 clock for pipeline to prime
            when PRIME_PIPE =>
                cnt <= cnt + 1;
                sc_bits <= SC(cnt);
                state <= MUL;
```

```vhdl
-- multiplication operation
when MUL =>
    -- update upper portion of accumulator with sum and
    -- carries from CSA above (offset carry by 1-bit)
    if(en_acc = '1') then
        acc(A_MSB downto A_MSB-53+1) <=
            signed(sout) +
            signed(cout(cout'high-1 downto 0)&'0');
    end if;
    -- right-shift upper accumulator into lower portion, but only
    -- when enabled
    if(shift_acc = '1') then
        acc(A_MSB-53 downto 0) <=
            acc(A_MSB-53+17 downto 17);
    end if;
    -- manage sequence code counter
    if(cnt = 0) then
        state <= MULX;
    elsif(cnt = MAX_CNT) then
        cnt <= 0;
        sc_bits <= SC(cnt);
    else
        cnt <= cnt + 1;
        sc_bits <= SC(cnt);
    end if;
when  MULX =>
    -- set rounding bits based on lower accumulator
    if(P_LSB > 0) then
        grs(2) <= acc(P_LSB-1);
    end if;
    if(P_LSB > 1) then
        grs(1) <= acc(P_LSB-2);
    end if;
    if(P_LSB > 2) then
        -- product is (+)
        if(acc(P_MSB) = '0') then -- product is (+)
            grs(0) <= '0'; -- initial
            for i in P_LSB-3 downto 0 loop
                if(acc(i) = '1') then
                    grs(0) <= '1';
                end if;
            end loop;
        -- product is (-)
        else
```

```vhdl
                        grs(0) <= '1'; -- initial
                        for i in P_LSB-3 downto 0 loop
                            if(acc(i) = '0') then
                                grs(0) <= '0';
                            end if;
                        end loop;
                    end if;
                end if;
            end if;
            -- normalize result back into accumulator
            acc(D_OVR downto 0) <=
                acc(P_MSB)&acc(P_MSB downto P_LSB);
            state <= ROUND;
            -- set overflow flag based in sign
            for i in A_MSB downto P_MSB loop
                if(acc(i) /= acc(P_MSB)) then
                    ovrflw <= '1';
                    busy <= '0';
                    state <= START_MUL;
                end if;
            end loop;
        -- rounding for multiply
        when ROUND =>
            -- round result to nearest even for positive
            if(ROUNDING = '1') then
                -- round to nearest even regardless of sign
                if(grs > 4 or (grs = 4 and acc(0) = '1')) then
                    acc(D_OVR downto 0) <= '0'&acc(D_MSB downto 0) + 1;
                end if;
            end if;
            state <= ROUND2;
        when ROUND2 =>
            -- check for overflow from rounding
            if(acc(D_OVR) /= acc(D_MSB)) then
                ovrflw <= '1';
            end if;
            busy <= '0';
            state <= START_MUL;

        when others =>
            state <= RESET;
    end case;

end if;
```

end process;

-- output signals
cmplt <= (not start) and (not busy);
product <= acc(D_MSB downto 0);

end RTL;

9 Other information on Multipliers

Names of individuals doing work on high performance multipliers include:

- Andrew Donald Booth
- Luigi Dadda
- Chris Wallace
- Peraris
- Baugh-Wooley

10 Addendum

The source code that follows supports conversions between real and fixed-point numbers, which exceed 32-bit limits.

```vhdl
--
--  File: ConversionPackage.vhd
--
library IEEE;
use ieee.std_logic_1164.all;
use ieee.numeric_std.all;
use work.all;

-- package header
package ConversionPackage is

    function RealToQmn(r: real; m,n: integer) return signed;
    function QmnToReal(qmn: signed; m,n: integer) return real;

end;

-- package body
package body ConversionPackage is

    --
    --  Convert real number to Qmn fixed-point number
    --
    --  m - number of integer bits excluding the sign (no limit)
    --  n - number of fractional bits (no limit)
    --
    function RealToQmn(r: real; m,n: integer) return signed is
    variable rx,tmp: real := 0.0;
    variable int: unsigned(m-1 downto 0) := (others=>'0');
    variable frac: unsigned(n-1 downto 0) := (others=>'0');
    variable qmn: signed((m+n) downto 0) := (others=>'0');
    begin
        -- convert to positive
        if(r < 0.0) then
            rx := (r * (-1.0));
        else
            rx := r;
        end if;
        -- compute integer portion
```

```
for i in m-1 downto 0 loop
    -- integer size limit
    if(i < 31) then
        -- subtract binary equivalent
        if((rx - real(2**i)) >= 0.0) then
            rx := rx - real(2**i);
            int(i) := '1';
        end if;
    -- multiply up limit by 2
    else
        tmp := real(2**30);
        for j in (m-1) downto 31 loop
            tmp := tmp * 2.0;
        end loop;
        -- subtract binary equivalent
        if((rx - tmp) >= 0.0) then
            rx := rx - tmp;
            int(i) := '1';
        end if;
    end if;
end loop;
-- compute fractional portion (remaining in rx)
for i in n-1 downto 0 loop
    -- integer size limit
    if((n-i) < 31) then
        -- subtract reciprocal
        tmp := (1.0/real(2**(n-i)));
        if((rx - tmp) >= 0.0) then
            rx := rx - tmp;
            frac(i) := '1';
        end if;
    -- divide down limit by 2
    else
        tmp := 1.0/real(2**30);
        for j in 31 to (n-i) loop
            tmp := tmp / 2.0;
        end loop;
        -- subtract reciprocal
        if((rx - tmp) >= 0.0) then
            rx := rx - tmp;
            frac(i) := '1';
        end if;
    end if;
end loop;
```

```vhdl
    -- construct final fixed-point number
    qmn := signed('0'&int&frac);
    if(r < 0.0) then
        qmn := (not qmn) + 1;
    end if;

    return(qmn);

end function;
--
-- Convert Qmn fixed-point number to real
--
-- m - number of integer bits excluding the sign (no limit)
-- n - number of fractional bits (no limit)
--
function QmnToReal(qmn: signed; m,n: integer) return real is
variable qmnx: signed(qmn'high downto qmn'low) := (others=>'0');
variable r,tmp: real := 0.0;
begin
    -- convert to positive number
    if(qmn(qmn'high) = '1') then
        qmnx := (not qmn) + 1;
    else
        qmnx := qmn;
    end if;
    -- compute integer portion
    if(m > 0) then
        -- add corresponding power of 2
        for i in qmnx'high-m to qmnx'high-1 loop
            if(qmnx(i) = '1') then
                -- integer size limit
                if((i-n) < 31) then
                    r := r + real(2**(i-n));
                -- multiply up limit by 2
                else
                    tmp := real(2**30);
                    for j in (i-n) downto 31 loop
                        tmp := tmp * 2.0;
                    end loop;
                    r := r + tmp;
                end if;
            end if;
        end loop;
    end if;
```

```
    -- compute fractional portion
    if(n > 0) then
        -- add corresponding power of two reciprocal
        for i in qmnx'high-1-m downto qmnx'low loop
            if(qmnx(i) = '1') then
                -- integer size limit
                if((n-i) < 31) then
                    r := r + (1.0/real(2**(n-i)));
                -- divide down limit by 2
                else
                    tmp := 1.0/real(2**30);
                    for j in 31 to (n-i) loop
                        tmp := tmp / 2.0;
                    end loop;
                    r := r + tmp;
                end if;
            end if;
        end loop;
    end if;
    -- restore sign
    if(qmn(qmn'high) = '1') then
        r := (r * (-1.0));
    end if;

    return(r);

  end function;

end;
```

www.ingramcontent.com/pod-product-compliance
Lightning Source LLC
Chambersburg PA
CBHW071805170526
45167CB00003B/1178